#TH!NKtweet

First Printing: June 2009

Paperback ISBN: 978-1-60773-044-6 (1-60773-044-8)
Place of Publication: Silicon Valley, California USA

Paperback Library of Congress Number: 2009929067

eBook ISBN: 978-1-60773-045-3 (1-60773-045-6)

Trademarks

Warning and Disclaimer

Advance Praise
(in alphabetical order)

"Treats for your brain in 140 characters."
@chrisgarrett

Author of ProBlogger

"#Th!nkTweet from @UpbeatNow takes smart, unconventional ideas and reduces them to the essentials. Highly recommended."
@chrisguillebeau

Blogger, The Art of Non-Conformity

"Think Tweets are too short to make you think? Raj's #Th!nkTweet will make you think again!"
@markmcguinness

Blogger, Wishful Thinking blog

"#Th!nkTweet = genius! ...a stellar job Twitter changed my life in 7/07: I see EVERYTHING in tweetable soundbites that adds value to others. "
@MariSmith

Social Media Expert, Relationship Marketing Specialist

Dedication

To my long-time hero **Tom Peters** who has made me Th!nk and Th!nk Again over these years.

Acknowledgments

First, thanks to Twitter for providing the inspiration to write this book.

Thanks to Chris Brogan *@chrisbrogan*, Joel Comm *@joelcomm*, Chris Garret @chrisgarrett, Chris Guillebeau *@chrisguillebeau*, Gerry Riskin @Riskin, Mark McGuinness *@markmcguinness* and Mari Smith *@MariSmith* for their kind words of praise.

My close friend Arun Nithyanandam *@StrategyNow* for all his insights.

Thanks to my long-time friend and publisher, Mitchell Levy *@HappyAbout* for being willing to experiment with this book.

Special thanks, as well, to Guy Kawasaki *@guykawasaki* for his twitterific foreword.

To Karine, Francis and the team at stresslimitdesign for their help with editing, as well as design.

To my wife Kavitha and our son Sumukh for providing the love and support needed to continue doing what I want to do.

Why I wrote this book?

It's a 24/7 world out there.

___

The success of Twitter has redefined how people pay attention, learn, collaborate and grow.

___

It's one Tweet at a time!

___

People want to read, learn and grow. But they don't have a lot of time to invest.

___

I created the #Th!nkTweet series as a solution!

___

Read, learn, think and grow with #Th!nkTweet.

___

All the very best!

___

Rajesh Setty *@UpbeatNow*

Bite-sized lessons for a fast paced world!

#Foreword

#Th!nkTweet is a cool little book filled with twinsights, twumor, and twinfluence of Twitter.

@guykawasaki
founder of Alltop.com

Bite-sized lessons for a fast paced world!

1

Networking 101 — If your goal is to ALWAYS give to your network, you will ALWAYS have enough to give.

2

Networking metric is NOT how you leverage your network but how much you CONTRIBUTE to the network (whatever be the medium.)

3

You are an "expert" when people who are qualified to make that assessment say so; NOT when you just claim it.

4

If you are the "signal," you don't have to complain about the noise. It's what will amplify your presence.

5

Mediocre help is everywhere. You can get it for less too.

6

Really good help may not be available even if you pay a premium. You have to earn it.

7

When you are REALLY good, people compete to work with you, since NOT working with you is a competitive disadvantage.

8

Social Media is about participation which is useless without contributions. So Social Media is all about preparation.

9

Direction is important.
If you're running fast
in the wrong direction,
you will reach the
wrong place — FAST!

DIRECTION FAIL !

Bite-sized lessons for a fast paced world!

10

If someone can copy your business by copying your actions, then there is a "structural" problem with your business.

11

I know one thing and that is "I don't know everything."

12

You steal an idea from a friend and you get ONLY one idea. Get that friend engaged and you get a GOLDMINE.

13

"My boss is the problem" is an excuse which is at the same level as "The dog ate my homework."

14

When it comes to relationships, you hit a home run when you stop keeping score.

16

The challenge is to leverage your PAST to be effective in the PRESENT while laying a foundation for your FUTURE.

15

You have a "problem" when you don't know the solution. You have a "bigger problem" when you don't know about the problem.

17

Stress is "trying to control what you know cannot be controlled" and forgetting to "engage in what you CAN control."

18

If people continue to listen to you, it could be that you are entertaining or enlightening. Don't confuse one for the other.

19

When you TRULY care for someone, his/her concern becomes your concern.

20

Make someone's dayEVERYDAY. It costs way less than you think.

Bite-sized lessons for a fast paced world!

21

If you don't know where you want to go, every place you go seems like a wrong place.

22

You can't solve the puzzle with only one puzzle piece in your hand.

23

Today is your last chance to do something about tomorrow.

24

Arrogance is a liability in the clothes of luxury.

25

"Who you are" is the lens through which people will read "what you write."

26

Justifying your addiction is sheer folly.

27

People try to invalidate a rule by stating an exception when exceptions are part of the rule.

28

One sign of healthy self-esteem is the ability to laugh at oneself.

29

Do you have any gaps in your organization? If yes, start filling them and you are on your way to becoming a leader.

30

Do you want a bigger slice of the pie? Then start with increasing the size of the pie.

31

You can talk anything (in your control) as long as you are willing to face the consequences (that may not be under your control).

32

How many of your close friends can DEPEND on you if they cease to be in their current positions?

Bite-sized lessons for a fast paced world!

33

What is the ONE habit that can totally change who you are for the better?

34

Fantasy is a dream without the action required to pursue it.

35

Be ready to provide significant value for free AND be ready to pay for significant value you receive.

36

Who is ONE person that can change who you are for the better? How can you be an OPPORTUNITY for him/her?

37

We value money as if we can NEVER get it back and we squander time as if there is an ENDLESS supply.

38

Who makes you THINK for the better? (Whatever be the answer, I hope you are in touch with that person daily).

39

The commitment you make to every promise you make to yourself will determine your self-esteem level.

40

What will be the title of a book written about you? What will be its subtitle? Why should someone read it?

41

Growing revenues is easy if you just have to do it on a spreadsheet.

42

Work and fun need NOT be mutually exclusive.

43

If it's easy for you to copy, remember that it will be easy for someone else to copy too.

44

You have three choices 1) complain 2) walk the extra mile or 3) create your own highway.

45

The results will be the reward for your client, and the journey towards those results will be the reward for you.

46

While you are traveling the road less traveled, remember to light the path for others behind you.

47

Want to be memorable? Be a catalyst for creating moments that are memorable in the lives of others.

48

Want to increase your power and influence? Expand the capacity of people in your network.

49

If you can't say "Thank God, It's Monday," think about changing your job OR changing yourself.

50

If nobody is listening when you are talking, shouting won't help.

51

When you say you have some "exciting news," it better be exciting news for the LISTENER.

52

Job search is done right when the right job comes searching for you.

53

You can pretend to care or you can care. In the same order, you can pretend to be fulfilled or be fulfilled.

54

Vividly imagining a fantasy won't make it a reality. You still need to work to make it happen.

55

Every "today" lived right will make every "tomorrow" much better.

56

Your power is directly proportional to the obligations you create with powerful people.

57

Every tweet can count if you want to make them count.

58

You might forget about something unpleasant you said. However, the listener might NEVER forget what was said.

59

The secret of success is to do what you love PLUS do what is required.

60

There are bad people, good people and those that pretend to be good people. C'est la vie.

Bite-sized lessons for a fast paced world!

61

No brownie points for walking the "extra mile" in the wrong direction.

62

Luck: The missing ingredient in your success and the secret ingredient in your neighbor's success.

63

The quality of your daily conversations will influence the rate of growth of your influence.

64

If you are "faking it" so that one day you can make it, you are missing the point. It's the journey that counts.

65

It is hard to "get ahead" if you are comfortable with just "moving along."

66

If you are adding "zero value" to a relationship, you might be an "opportunity cost" for the other person.

67

You are missing the point if the listener missed what you said but didn't miss much.

68

Sometimes people get very busy worrying about how busy they are.

69

The best referral is one that you get
without you asking for it.

70

When you become somebody, it's
easy to forget those that were with
you when you were nobody.

71

Mediocrity has no defined standard. Once someone raises the bar, whatever is below that falls into mediocrity.

72

Invest in yourself. As you keep increasing your capacity to give more, it keeps costing you less to give more.

73

A quick short-term victory (with a long-term loss of trust): Take credit for the work of someone else.

74

When you trust someone but put them to the test everyday do you really trust them?

75

If you have a "really" good solution and you can wait long enough, you will find a "problem" for that solution.

76

"Above average" people think there is a long way to go while "average" people think they are "above average."

77

The ultimate fantasy: To expect spectacular results from mediocre efforts.

Bite-sized lessons for a fast paced world!

78

Everyone knows that one can't drive looking at the rear-view mirror but most won't follow that rule in their lives.

79

It is easier to come up with an excuse to NOT pursue your dream than to put the effort into pursuing it.

80

If you are NOT in the middle of something important, you might be in the middle of a mess.

81

1. New media is talking about "old media being dead." 2. Old media is talking about "new media" to stay alive.

Bite-sized lessons for a fast paced world!

TEAM WORK!

82

Whether you like it or not, you need a certain level of maturity to notice and accept "good help."

83

Your ego will "protect" you from new learning.

84

You can always blame it on **luck**.

85

What is the ONE thing that you can do right now that will make the people that you respect proud?

86

The ultimate excuse for not doing anything: Not being sure of what you want in life.

87

An hour spent on what "you can't control" is an hour you could have invested in what "you can control."

88

"Scarce and Valuable" commands a premium. Creating "False Scarcity" will raise questions about your credibility.

89

It's hard to find a solution when you are part of the problem.

90

Alternate routes to your goal may exist but they may show up only when you are halfway through your journey.

91

You can invest the next hour settling scores OR scoring new goals. Your choice.

92

Thinking about doing something important is always easier than actually doing it. But doing it is where all the fun is.

93

"Good Advice" may be expensive but it's cheaper than the price you pay for not taking it.

94

If the only way you can prove your point is by proving the other person wrong, think again!

95

People who can take care of themselves are open to help and people who need help don't think they do.

96

Good moods may not
open new opportunities
but bad moods
will close existing
opportunities.

97

You need help when you don't know everything. You need LOTS OF HELP when you think you know everything.

98

The best help you can get is one which helps you "discover and solve" problems that you didn't know existed.

99

If "work-in-progress" worries you, imagine a day when there is no work to make any progress.

100

The ultimate appreciation: Accepting someone just the way he/she is.

Bite-sized lessons for a fast paced world!

WHERE ARE YOU ?

101

If you disappear from
the world today, how
big of a void will it
create in all your
networks?

Bite-sized lessons for a fast paced world!

102

You can have "doing nothing" as an option as long as you are comfortable with "having nothing" as a consequence.

103

The joy lies in helping someone achieve something that they could not have achieved on their own.

104

Communication includes transmission and reception. Good communicators take responsibility for both.

105

One gets "hurt" when ignored. Rather than that, why not work towards "being" that someone who cannot be ignored?

106

Keep thinking of things you hate and you won't have time to discover things you love.

107

You want help from other people?
Make it easy (low-cost) for them to
help you.

108

When you treat every rejection as a
feedback and DO something about it,
life will get easier.

109

Life is an eternal "work-in-progress" project.

110

You cannot "stand out" by conforming. Nonconformance alone won't make you "stand out" either.

111

140 characters is a LOT when you don't have anything to say.

112

Everybody thinks they are "above average" proving that a majority of the people have problem with maths.

113

We accept defeat in our mind first and then find a "good" excuse to explain to ourselves and to others.

114

A really good "excuse" is still an "excuse."

115

Don't try to get an
award for the best
excuse. There is too
much competition in
that game.

116

I feel sorry for "luck." Most people don't give "luck" the credit it truly deserves.

117

If people listen to good advice and do nothing about it, they don't want advice, they want entertainment.

118

Do you stretch your customers' imagination beyond their imagination?

119

You can't lose what you don't have.

120

Since who you are is reflected in all your actions, one way to change your actions is to change "who you are."

121

Why should anyone believe you more than you believe in yourself?

122

A keyboard is mightier than the sword. (ONLY if you use it well though!)

123

Whether you like it or not, you are playing a high-stakes game (remember, you get to live only ONCE).

124

What was your proudest moment last year? How many such moments will you create this year?

125

What will you complete today?

126

Who is taking you for granted?

127

Nobody won an award for "standing on the sidelines."

128

Two related questions:1) Do you deserve what you want? AND 2) Have you earned what you deserve?

129

If it's broken you have to fix it. If it's not broken, your competition is looking at it, so "fix it" anyway.

130

You can make a quick buck ONLY if you present a compelling offer to people who want to make a quick buck.

131

Greater the challenge, sweeter the victory.

132

It will wok out (provided you do your part).

133

Enemy #1 for any startup (or for anyone trying to initiate change): "Status Quo."

134

If you give out only the very best, sooner than later, the world will reciprocate.

Bite-sized lessons for a fast paced world!

135

The best call you can make tomorrow is a call to your "calling."

136

The tragedy is not the lessons you missed learning but the lessons you missed noticing.

137

A quick way to get annoyed: Try helping someone who does NOT want to be helped.

138

If you magically put in the effort in the right places, you will magically see the results you deserve.

Bite-sized lessons for a fast paced world!

139

The classic dilemma: 1) Want to grow AND 2) Want to continue to operate in the comfort zone.

140

Today is your chance to make the "future of yesterday" and the "memories of tomorrow" special.

www.ingramcontent.com/pod-product-compliance
Lightning Source LLC
Chambersburg PA
CBHW060553100426
42742CB00013B/2542